WOOD BURNERS

DANIEL MIHALYO

——

PRINCETON ARCHITECTURAL PRESS

Published by:
Princeton Architectural Press
37 East 7th Street
New York, NY 10003

Edited and designed by
Mark Lamster

Printed and bound in China

For a free catalog of books published by Princeton Architectural
Press, call toll free 1.800.722.6657

Visit Princeton Architectural Press on the World WIde Web at
http://www.papress.com

Library of Congress Cataloging-in-Publication Data

Mihalyo, Daniel, 1970–
        Wood Burners / Daniel Mihalyo
               p.   cm.
        Includes bibliographical references (p. 118).
        ISBN 1-56898-104-X (alk. paper)
        1. Incinerators—Northwest, Pacific—Design and
        Construction. 2. Sawmills—Waste disposal—
        Northwest, Pacific.   I Title.
        TD796.M54 1997
        674' .2—dc21   97-976
                                            CIP

For Katsumi, Clausen, and O'Riley
simple is best—simple is deep

CONTENTS

Wood burner, Fruitvale, British Columbia.

Foreword

In 1992, when Daniel first sent me photographs of his study of the sawmill wood burners of the Pacific Northwest, I was stunned by the poetic intensity of his interior views. Was it the magic of the glistening light slicing into black conical volumes? Was it the melancholic sense that all of these magnificent steel constructions were being torn down and would be forever lost?

Could it have been my childhood memory of being mesmerized by the big rust covered burner on the road from Bremerton to Gorst in Washington, with sparks glowing like fireflies from its top in the winter twilight? Was it a deep conviction that I had architectural experiences at the age of five or six, even though I lived in a small town without a famous building or architect?

Soaked in rust, burners are pure constructions joined by steel skins topped with wire. They have a tangible presence expressed through material purity. Their simple geometry is conveyed with elegance; a cone remains an elemental form in the cube and sphere based language of geometry.

Wood burners unburdened of their function fall into silence. They find a new presence at once poetic, ordinary, and strange. Consider the burner in relation to Ludwig Wittgenstein's poetics of everyday life. Once the functional trivialities of daily use are stripped away, a new understanding of space, geometry, and light, is established. Forms of construction that were never associated with words like *beautiful* and *good* are reappraised. What is poetic? Following Roman Jakobson, the poetic is "present when the word is felt as a word and not a mere representation of the object being named . . . when words['] . . . external and inner form acquire a weight and value of their own instead of referring indifferently to reality."

With their burning function eliminated, when one can walk inside, these are astonishing constructions. The phenomenally and haptically intense combination of light and space that lies within Rome's Pantheon is echoed in the wood burner, but in different terms. The arch of the sun rotates through a blackened steel cone with a new silence and intensity. Measuring the seasons in altitude and angle like giant calendars, these empty, sooted cones trap time. In their shadows, the light of the moon or the sun can be stopped in a photograph. The weight of time goes on . . .

Steven Holl

Wood burner, Willow Ranch, California.

Archaelogy

It first appeared at the peripheral limit of sight, a blip on the horizon with a curious power of attraction. Eventually, it resolved into a conical structure, a matte-black mass overwhelming the surrounding scrub, soil, and grass.

Save for a concrete foundation, the structure consisted of uncoated steel that shamelessly displayed the beating, gouging, scorching, and rusting of decades of exposure and use. Unencumbered by decorative detail, there was nothing to indicate whether it was a machine in itself or simply a piece of architecture sheltering one.

In the course of circumnavigating its perimeter, an opening in the foundation appeared that was just large enough for a hunched-over person to enter. Passing through the sheet steel membrane and into the core, the structure's illusion of solidity was burst, resulting in a bewildering spatial experience. Within the core, a lack of context and an unfamiliar landscape produced a sublime and transcendental quality. Such an encounter with wrapping space and haunting chiaroscuro has few analogs in the history of architecture.

In the moments that followed, the eyes calibrated to the darkness of the interior of the vessel, revealing the sources of the structure's phenomenal displays of light. In contrast to the full sunlight on the exterior, small amounts of light twinkled and cut through the mechanical openings, thermal fractures, and shotgun spatter that punctured the walls. High overhead, a large, circular opening provided a link to the sky. Light fell through this aperture, spilling against the walls, casting web-like patterns against the vertical surfaces. Through the course of the day, great arcs of light formed and swept about in elliptical pools, eventually evaporating into the extremities of the cone.

Taxonomy

The sheer mass of history that has accumulated in the brief one hundred fifty years of Western settlement in the Pacific Northwest is already more than one can hope to digest in a lifetime. By focusing on the metamorphosis of a single typology, however, one can explore this vast deposit of ideas and meaning to a degree not possible in a study of broader scope. Although this examination of the sawmill wood burner and its history is nominally "architectural," it also seeks to illuminate issues of cultural heritage and resource conservation insofar as they relate back to the subject of wood burning.

The foundations of industrial forestry in the Pacific Northwest date to 1827 with the partnership of John McLoughlin and William H. Crate. With financial backing from the Hudson Bay Company, they imported enough machinery from London to set up the first water-powered sawmill on the West Coast, approximately six miles east of Fort Vancouver, Washington.[1] The lumber they produced

Sawmill in operation at Quesnel, British Columbia, 1996.

went to the Hawaiian Islands, where it was used to help establish plantation agriculture, and to other small markets in California. In the years that followed, sawmills sprouted up at commercial centers up and down the West Coast. By 1844, the market for their lumber products had expanded to include Australia and China.[2]

The 1848 discovery of gold at Sutter's Mill, California, in the foothills of the Sierra mountains, brought thousands of Eastern transplants to the region in search of opportunity and fortune, generating a vast market for sawn lumber. With the forests east of the Mississippi nearing exhaustion, this migration also brought the financial resources of Eastern timber barons who sought to capitalize on the situation. In addition to supplying the materials for the rebuilding of San Francisco after seven disastrous fires, the expansion of major transcontinental railroads and shipping lanes meant that by the 1880s the Douglas fir lumbering industry had expanded its markets to include all of North and South America and the entire Pacific Rim.[3] These were the glory days of lumbering, with plentiful jobs and a limitless resource slated for prosperous exploitation. Government experts of the era earnestly believed that the forests of the Pacific Northwest would satisfy the nation's appetite for lumber in perpetuity.

The introduction of steam power and higher strength steels for use in the bandsaws used to cut lumber had an enormous impact on the production capacities of even the smallest mills. With these innovations came astounding volumes of waste composed of sawdust, slabs, chips, planer shavings, bark, edgings, remnants, and defects. Statistical evidence concerning the actual volume of wood waste generated at the average sawmill during this period is, unfortunately, difficult to interpret. Nevertheless, even by the 1950s only 30 to 40 percent of every logged tree was utilized, and this included the use of sawmill waste in the manufacture of paper, plastics, rayon, turpentine, broom handles, pressed fire logs, and particle board. In earlier periods these percentages were undoubtedly even lower.

At the traditional sawmill, a small portion of scrap was used to fuel the steam boilers that powered the mill complex. But in one day's production a small mountain of debris could accumulate, and disposal proved to be a con-

Fig. 1. Waste piling up at Eagle Creek (Oregon) Sawmill, circa 1867. Photo courtesy Oregon Historical Society.

stant nuisance. Often, this waste was incinerated in open pits. More commonly, it was dispersed by floating it downriver or by distributing it into the lowland areas that surrounded the mill (fig. 1). In fact, several acres of Seattle are built atop fifteen feet of just this type of residue, a byproduct of the city's founding sawmill.

Mills lacking adequate disposal facilities could be literally buried in piles of flammable sawdust and kindling. Fire struck frequently, and those who were uninsured lost everything. Lower insurance rates, however, could be had by those mills exhibiting a safe means of disposal. Early historic photographs suggest that open-pit incineration eventually gave way to burning in steel lined enclosures, with waste delivered by sloping conveyor belts.

In 1888, Muskegon Boiler Works of Michigan was issued the first patent for a structure specifically designed for burning sawmill waste (fig. 2). In an effort to accom-

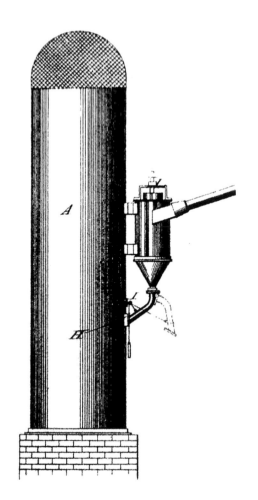

Fig. 2 (left). Illustration of wood burner from 1888 patent application by Thomas Munroe and Hugh Park of Muskegon, Michigan.

Fig. 3 (above). "Water jacket" wood burner in La Grande, Oregon. Photo Courtesy Oregon Historical Society.

Fig. 4 (below). Advertisement from 1916 for the "air-cooled" wood burner produced by Colby Engineering of Portland, Oregon.

There is a Vast *Difference* in Burners

modate the needs of new sawmills, Muskegon was soon fabricating a variety of wood burner designs that resembled large-diameter, steel smokestacks. Each burner was designed with dimensions tailored to the type and anticipated volume of wood waste to be burned. Frequently, diameters reached forty-eight feet with heights in excess of 140 feet. The burner consisted of a concrete or stone foundation upon which rested a steel shell for general support and a thick refractory lining of fire brick. These burners were commonly pre-fabricated at Muskegon's Michigan plant and then, like other components for the new sawmills, shipped via railroad to the mill site along with a fabrication team. The success of this venture inspired numerous other fabricators to knock-off Muskegon's designs and vie for their market.

One step ahead of the competition, in 1908 Muskegon unveiled the "water jacket" wood burner (fig. 3). The new design incorporated a second layer of boiler plate that was separated from the inner lining by eighteen inches of water. This offered two attractive advantages over their original design. First, it kept the temperature of the inner lining of boilerplate in the range of 220 degrees Fahrenheit, the boiling point of water. Considerably below the slumping temperature of steel, this substantially reduced the required thickness of the expensive refractory interior lining. Second, the water between these two layers was ingeniously preheated for use in the mill's adjacent powerhouse boilers. The substantially higher cost for this design, however, made it available to only the largest and most profitable mills (fig. 13).

The basic, cylindrical wood burner dominated the sawmill industry from the mid-1800s through World War I. By the end of this period, sawmills peppered the landscape of the burgeoning Northwest in great numbers. The constant and heavy demand for lumber, compounded by technological improvements in all phases of the foresting industry, created ever greater quantities of sawmill waste. Although every mill needed some method of wood waste disposal, the costly pre-fabricated wood burner remained an impossible luxury for many smaller operations.

Seeking to capitalize on the demand for a less costly incinerator, Colby Engineering of Portland, Oregon introduced a completely new type of wood burner in 1916 that relied on a patented, "air-cooled" design (fig. 4). In an effort to eliminate the expensive refractory lining, Colby's burner flared out conically at the base in order to keep flames away from the shell. As with Muskegon's "water jacket" burner, a second steel shell was added and separated from the first shell by a riveted truss. At the burner's base, air—instead of water—was introduced into the space between the two shells and was then drawn upward by the natural draft of the flames. In theory, this air-flow would keep the shell cool. This allowed for the use of relatively thin-gauge plate steel cladding, as opposed to the thick boilerplate required by previous burners, even further reducing costs. The uppermost section of these burners retained the cylindrical shape of earlier burners, while an elaborate foundation housed tunnels supplying oxygen to the burning embers. These tunnels also allowed workers to periodically enter and clean the burner's ash pit.

This new design received excellent coverage in trade journals, and its modest price made it widely attractive to mill owners. Soon after the introduction of Colby's air-cooled burner, however, an inspired burst of applications arrived at the patent office, claiming ever greater reductions in material and construction costs. The rapidly expanding market for wood waste incinerators created its own log-jam of entrepreneurs and inventors.

In 1917, Frank Hopkins, owner of the Seattle Boiler Works, made a patent application for the first truly lightweight wood burner design (fig 5). It relied on a single layer of thin-gauge plate steel cladding arranged in a simple conical form with spot-footings at each vertical support. Air was introduced by means of adjustable dampers on the lower circumference of the shelled surface. Practical and affordable—almost crude—this design proved to be enormously popular with mill owners throughout the Pacific Northwest.

Ironically, the improvements patented by Hopkins simplified the construction process to such a degree that resourceful mill owners began fabricating their own burners. This resulted in a flurry of ameliorations and variations to the typology that went unnoted in the patent register. By the mid-1920s, the wood burners appearing at mill sites

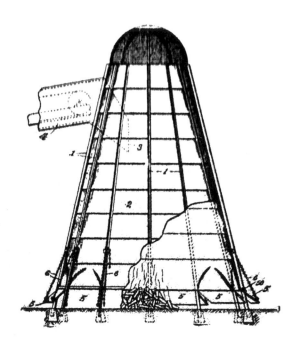

Fig. 5. Illustration of wood burner from 1917 patent application by Frank Hopkins of the Seattle Boiler Works.

and quantities of waste to be incinerated, resulted in a bewildering assortment of variations on the standard burner typologies. Nevertheless, by the mid-1940s the dozens of independent fabricators had come to a silent consensus on the basic form of the structure. This convergence can be attributed to a pair of rules of thumb widely adopted by those erecting burners. The first of these held that the diameter of the base equal the altitude of the frustum (the distance between the base and cap diameters); the second that the diameter of the cap should be half that of the base.

Advances in basic burner design were accompanied by improvements to the fabrication process. Construction of the basic cylindrical shell had been reduced to the repetitive fabrication of small segments of the larger conical whole. All that this required was an accurately sized template and a rolling machine to bend the steel plate. Once the pieces were made and the concrete foundation poured, the entire shell could be assembled, in the course of an afternoon, with the help of a crane hoist (fig. 6). By 1950, the relative ease of construction made it possible for any of the more than two thousand mills operating in the Pacific Northwest to acquire a burner on short demand. If production increased, mills could simply add burners as needed. In fact, it was not unusual for prosperous mills to operate three of four burners simultaneously,

For a brief period shortly after the bombing of Pearl Harbor, the use of wood burners became a national safety issue. Defense wardens up and down the Pacific Coast expressed concern that the glow and sparks emanating from burner domes lit up the coastline "like the biggest Christmas tree of all time."[4] During the frequent wartime blackouts, all coastal wood burners were ordered extinguished. Relighting the drenched piles of sawdust within the burners, however, proved difficult, requiring a substantial quantity of crude oil that was then in short supply. This led to an impasse, as the military was then placing huge lumber orders to support the war effort. A creative compromise solution allowed mill owners to install sprinkler systems in their incinerators that would lightly douse flames in the event of an air-raid.[5] Today, one can still find the occasional sprinkler pipe running alongside the shell of a coastal wood burner.

throughout the region had incorporated several additional design improvements. The exterior application of trussed compression rings rapidly became a standard fixture in burner design, dramatically improving the general strength of the structure in the face of the extreme temperatures regularly produced inside. Additionally, it was discovered that if air was tangentially introduced into the burning pile of embers within, the ash and ignited particles would be swept about the interior chamber in a spiral motion. Commonly called "cyclone action," this prolonged the burning process and more effectively incinerated escaping particulate matter. The exact arrangement of these improvements varied with every wood burner as individual fabricators sought the optimal conditions for the type and quantity of waste to be incinerated.

The multiplicity of fabrication techniques, budgets, and building traditions, combined with the differing types

Fig. 6. Segmental construction of wood burner in Springfield, Oregon circa 1920. Photo courtesy Oregon Historical Society.

Fig. 7. Gas tank inverted and converted to use as a wood burner near Forks, Washington.

During the post-war building boom in the United States, wood stud framing emerged as the predominant construction method for single family homes. A lucrative market coupled with the advent of inexpensive chainsaws, diesel generators, and portable sawmills spawned a new breed of woodsman known colloquially as the "gyppo." These resourceful and independent individuals performed every task from felling trees to bucking, yarding, and even milling lumber. Gyppos operated with a minimum of resources and, unable to buy it new, frequently fabricated their own machinery. They quickly found that discarded gas tanks could serve as adequate and inexpensive stand-ins for the costly, pre-manufactured wood burner. This simple conversion entailed standing the tanks on their ends, installing a few courses of fire brick on the interior, cutting a hole in the top, and supplying air intake holes at the base (fig. 7). The gyppos, who existed in great numbers during the prosperous decades following the war, eventually fell prey to larger competitors and successive housing slumps.[6]

The post-war decades also witnessed the large-scale development of industries created specifically to make economic use of wood waste. Despite the enormous success of products like particle board, however, and even with the advantage of new, highly efficient milling equipment, a tremendous volume of wood waste still required incineration. By the 1960s, the forest industry was confronted with a public whose consciousness of air pollution problems had reached critical mass. This resistance eventually manifested itself in the far-reaching Clean Air Act of 1970, which passed through both houses of Congress with only one dissenting vote.[7] Considered a landmark of environmental legislation, the Clean Air Act created benchmarks and regulations requiring dramatic improvements in all polluting

Fig. 8. Olivine type wood burner, Lumby, British Columbia.

industries. As the most potent symbol of air pollution in the forest industry, the wood burner became the first target for improving visible emissions at sawmills.

As early as 1967, the Oregon State Legislature, sensing the rise in public pressure to curb pollution, directed the Forest Research Laboratory—a department in the School of Forestry at Oregon State University—to develop potential solutions for the problem of wood burner emissions. After extensive testing, the team concluded that smoke levels could be dramatically decreased by increasing combustion temperature and adjusting the way oxygen was introduced into the burner. By installing large semicircular lids that were coupled to the readings of a temperature gauge, it was determined that the interior temperature could be maintained at an optimal eight hundred degrees Fahrenheit. Additionally, numerous underfire and overfire supply vents, powered by large fan units, could increase the velocity, quantity, and direction of the oxygen used to power the burner.[8] These design retrofits were tested on a full-scale burner, where they proved to substantially reduce visible emissions, provided that the waste material itself was consistent and dry.[9] The cost for these alterations, however, was high.

After the new laws came into effect, mill operators were required to invest in these expensive improvements in order to comply with regulations. Either unwilling or financially unable to do so, most mills abandoned burning altogether. Those that had economically viable access to a market for their waste—a paper mill or particle board plant, for example—took advantage of this opportunity. But the economics of the building industry, the fluctuations in paper pulp prices, and the high costs of shipping—due to the Arab oil embargo—often precluded mills from realizing a profit from these secondary sources. Moreover, even by 1970 only 50 percent of every tree cut was utilized.[10] With no alternative method of disposal for their mounding waste, many mills were forced to continue burning under the increasingly restrictive—and expensive—environmental standards.

The olivine burner, a more efficient type of incinerator, was also introduced during this period (fig. 8). Consisting of a refractory chamber made of pre-cast panels of olivine—a complex silicate of iron and magnesium developed to withstand high temperatures—these cylindrical burners were equipped with high powered fan units and constricting caps that allowed the burners to withstand and control temperatures in excess of fifteen hundred degrees Fahrenheit. Although olivine burners were exceptionally efficient, their extremely high price placed them beyond the reach of most mill operators, resulting in little effect on the industry.

With the end of the oil crisis of the 1970s, and with a continually improving system of roadways, most mills were able to profit by trucking their waste to secondary markets, often several hundred miles away. And even if trucking was not entirely profitable, mills resorted to it anyway—forsaking incineration—in order to avoid public resistance and federal monitoring. It was not uncommon, in the mid-1970s, to see burners being dismantled at mill sites by owners seeking to avoid even the suspicion that they were polluting (fig. 9). By the early 1980s, the American forest products industry had effectively abandoned the practice of wood burning and proudly boasted of "full wood utilization." Despite these lavish claims, though, one can still find a remote family owned mill, hidden at the end of a long, dirt road, that will occasionally fire up its wood burner to dispose of the worst of its waste.

In Western Canada, however, which is often said to be twenty-five years behind the United States in both environmental legislation and secondary wood products infrastructure, the wood burner remains a ubiquitous presence. This has allowed the typology to experience another generation of functional improvements.

Foreseeing the gradual northern migration of public resistance to air pollution, several Canadian fabricators responded by creating their own versions of the "smokeless" wood burner (fig. 10). The first of these designs appeared in the late 1970s, and they eventually became the only viable alternative to the increasingly stringent air quality standards then beginning to effect Canadian sawmills near urban centers.

The Canadian smokeless burner differs from its American counterpart by using four upward acting damper lids at its cap (as opposed to two downward acting lids on American designs), thereby realizing several significant improvements. First, these lids open in the direction of the

Fig. 9. Demolition of wood burner in Springfield, Oregon, circa 1970. Photo courtesy Oregon Historical Society.

Fig. 10. Canadian smokeless type wood burner, Terrace, British Columbia.

swiftly rising eight hundred to twelve hundred degree Fahrenheit air current. This creates less turbulence and therefore more efficiently directs sparks and hot air out of the burner. Moreover, the fact that these damper lids open away from instead of into the flames naturally contributes to a much lower frequency of thermal damage.

The earliest versions of this design were notable for the absence of a domed, spark arresting screen at the cap of the burner. In theory, the increased temperature at which these incinerators operated should have completely burned all embers before they could escape. This theory, however, only worked in practice when the burner operated under "ideal conditions," and ideal conditions were elusive in reality due to damp wood, dirt, and otherwise impure waste. More often than not, smoke and embers simply poured out of these burners, eventually leading to the reintroduction of the spark arresting screens.

Increasing pressure to run smokeless has forced many mill owners to run their burners at higher temperatures than those for which they were originally designed to withstand. Consequently, the steel cladding is frequently deteriorated due to extreme and persistent expansion and contraction. The first signs of this deterioration are the severe oil-canning of the shell followed by the development of small cracks and fissures in the plates. The cumulative air-leakage from these perforations adversely effects the performance of a burner, often requiring expensive maintenance or replacement. Conversely, this phenomenon is also responsible for the spectacular formations of light found on the interiors of many burners.

Final termination for Canadian wood burner permits was slated for late 1996. However, several large mills reached an agreement with Canadian environmental agencies allowing them to consolidate their waste for common

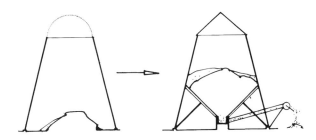

Fig. 12. Diagram of wood burner conversion to chip hopper.

Fig. 11. Wood burner converted to chip hopper, Prince George, British Columbia.

incineration through 1997. This will likely spell the end for the wood burner typology as it is known in North America. Nevertheless, the multinational forest industry is undoubtedly installing wood burners at mills recently constructed in the remote forests of South America, Southeast Asia, and the former Soviet republics.

Reuse
The wood burner's construction is simple, consists of primary components, and is detailed without contrivance in a manner anticipating persistent, heavy use. Little is custom made, difficult to replace, or fussy in design. These characteristics have allowed wood burners to be readily adapted to secondary uses—fortuitously, given the obviation of their intended function. Generally, these reuses have been utilitarian. Others, however, range from the creative to the whimsical to the truly bizarre.

The most common secondary use is the conversion of the burner to a crude wood chip or sawdust hopper (figs. 11 & 12). This is accomplished by fabricating a sloping trough that funnels the wood waste into the burner, from which it is eventually removed by a belt driven conveyer. More primatively, a burner may simply be filled to the brim with wood chips and then periodically emptied by a bulldozer. Evidence of this type of reuse is visible in several of the exterior plates in this volume, where wood chips can be seen spilling out and about the bases of individual burners (see plates on pages 101 & 103, for example).

Bonners Ferry, in the Idaho panhandle, was a prosperous mill town at the turn of the century (fig. 13). Eventually, however, its pine-lumber mill closed from lack of timber and fell into disrepair. For several decades following the closure, the wood burner was used as a grain elevator, allowing it to remain intact while the rest of the mill was gradually demolished. Eventually, the grain elevator venture also failed and the site, complete with wood burner, was purchased by a local entrepreneur who built an elaborate multi-story home within the thirty foot diameter core. Although it has been received with trepidation by a skeptical community, the unobstructed view from the top storey is excellent, and a more solidly built structure cannot be found.

North of Missoula, Montana, in the town of Ravalli, a life-size reproduction of a New Holland windmill has been constructed around the shell of a wood burner (fig. 14). This unique conversion included the installation of a fully

Fig. 13. Panoramic view of mill site at Bonners Ferry, Idaho circa 1923. The smoking, water jacket wood burner is located on the far left. Photo courtesy Bruce Teter.

Fig. 14. Wood burner converted to windmill/motel in Ravalli, Montana.

Fig. 15. Wood burner in junkyard near Bend, Oregon.

functioning motel complete with bar, spiral staircase, honeymoon suite, red shag carpet, and wrap-around balcony. The owner had dismantled the mill after retiring (he had been the mill's operator) in order to devote his golden years to fulfilling a dream of becoming an inn-keeper. Sadly, he died before receiving his first guest.

Perhaps the most eccentric of all wood burner conversions, however, can be found north of Bend, Oregon, on a former mill site now surrounded by barbed wire and patrolled by a guard dog. There, in the middle of a vast graveyard of car remnants, lies the only remaining vestige of the mill; a large, rusty wood burner. Encircling the burner's shell is a system of racks that displays the junkyard's best preserved automotive front ends (fig. 18). Commanding attention from the highway, it is a poignant—if odd—metaphor for the American recycling movement in the twentieth century.

Burning

The flames have disappeared, the red coals have cooled, and ash now endlessly circles so many dark chambers. Meanwhile, the abiding conclusions as to why these vessels have been forsaken remain contradictory. Those who claim to know about the fate of the wood burner may allege that they were outlawed by restrictive environmental legislation. Others might explain how developing technology in the forest and secondary wood products industries outmoded the need to burn wood waste. One might then conclude that we have progressed beyond our polluting and wasteful past.

While this reasoning may contain splinters of truth, it is surely the case that change in the aggressive forest industry would not have occurred without economic reason. With the inevitable depletions of the last great forests in North America, the forest industry reacted by adopting progressively more efficient methods to make what little was left last slightly longer. The wood burner was gradually abandoned not out of a genuine concern for air quality standards nor due to the benign cultivation of efficiency in order to better sustain ancient forests, but rather from fear of impending financial collapse. In short, by 1970 timber resources were so rapidly approaching total depletion that there remained no alternative for a profitable future other than the creative use of the industry's own excrement. What was formerly considered waste, refuse, and scrap—not even worth the cost of giving away—was ingeniously transformed into a financial motherload. The system of wood building that has consequently developed in the late twentieth century is composed of a bewildering array of engineered composite materials that might otherwise have been incinerated. Ironically, these secondary wood products are so heavily dependent on petrochemical adhesives, coatings, and preservatives that they eclipse any pretense of ecological progressiveness.

The burner is a casualty of economy and efficiency, the twin principles that made its existence possible in the first place. In turn, it has been replaced by an ever expanding collection of building materials that share in the further degradation of our ecology. In this light, it would be mistaken to look nostalgically upon the wood burner and imagine that it is a part of a more wholesome past. It remains a rather sinister fact that the wood burner indiscriminately devoured some 60 percent of all timber harvested through the 1970s in the Pacific Northwest and other previously forested regions of North America. Nevertheless, the wood burner holds a remarkable place in our history, and the knowledge of its history can be of real value to those earnestly seeking to transcend our wasteful past.

*　　*　　*

Although this history has been presented using the tools of architectural representation, it is not suggested here that the wood burner is architecture. It is a primitive machine. More accurately, it is an oversized vessel in which the transformation of materials occurs, not unlike a blast furnace or a ceramic kiln. This is not to sat that there are not architectural lessons to be gleaned by carefully studying this structure. The wood burner's often extraordinary spatial qualities, for example, can most certainly be reinterpreted to inform the making of architecture in the future.

Notes

1. Lewis L. McArthur, "Industrial Building," in *Space, Style and Structure,* Thomas Vaughn, ed. (Portland: Oregon Historical Society, 1964), 161.

2. Thomas R. Cox, *A History of Pacific Coast Lumber Industry to 1900* (Seattle: University of Washington Press, 1974), 1–22.

3. McArthur, "Industrial Building," 161–164.

4. Ellis Lucia, *The Big Woods* (Garden City: Doubleday & Co. Inc., 1975), 30-31.

5. Ibid.

6. Robert Leo Heilman, "Death of a Gyppo," in *Oregon Quarterly* (Eugene: University of Oregon, 1996), 20–23.

7. U.S. Congress, "Clean Air Act" (91st Congress, second session), *Congressional Quarterly,* vol. 26 (Washington, D.C.: Congressional Quarterly, Inc., 1970), 472–486.

8. George H. Atherton, Russell W. Bonlie, Stanley E. Corder, and Paul E. Hyde, "Wood and Bark Residue Disposal in Wigwam Burners," *Bulletin* 11, paper 720 (Forest Research Laboratory, Oregon State University, March 1970).

9. "Visible emissions," are widely accepted as the basis for nearly all air-pollution calculations. They are measured on the Ringelman Chart, a subjective method that rates emissions according to the light obscuration of a plume of smoke. The chart works on a scale of zero-to-five, with zero being the equivalent of clear gas and five a cloud of thick black soot. Wood burners operating under optimal conditions with the necessary pollution retrofit devices can achieve a zero rating.

10. Atherton, et al., "Wood and Bark Residue Disposal," 5.

TYPOLOGIES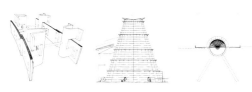

The Genealogy of the Wood Burner

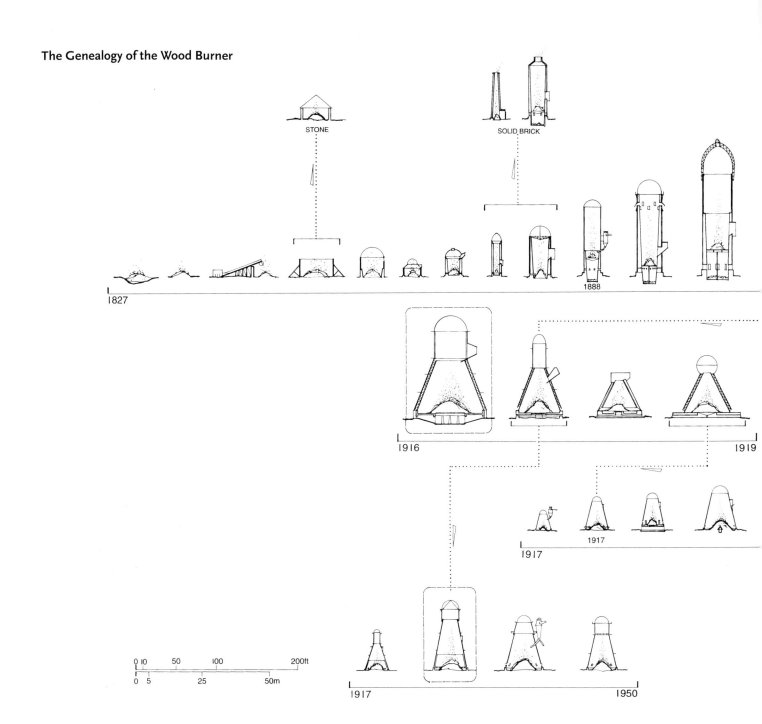

STONE

SOLID BRICK

1827

1888

1916

1919

1917

1917

1917

1917

1950

0 10   50   100   200ft
0 5   25   50m

24

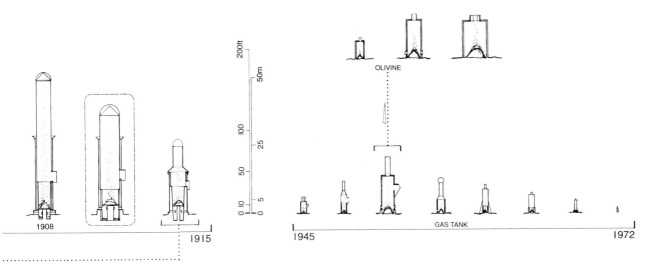

OLIVINE

1908   1915   1945   GAS TANK   1972

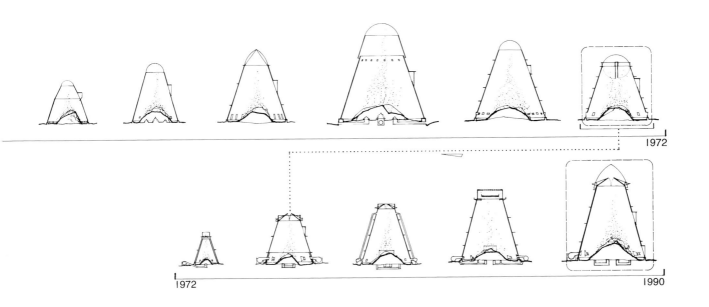

1972

1972   1990

Cylindrical & "Water Jacket" Burners

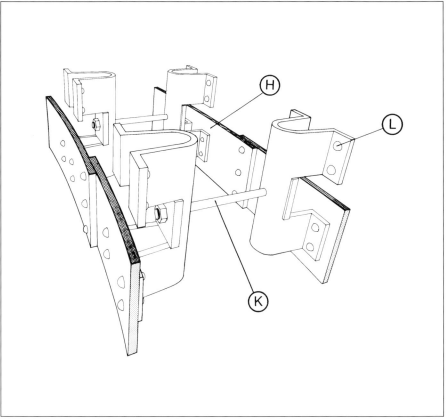

This, the earliest wood burner type, was little more than a large-diameter steel smoke-stack with a thick refractory lining that ran continuously up the interior sidewall. In fact, construction of cylindrical type burners was not limited to the Pacific Northwest; Michigan's Muskegon Boiler Works was the preeminent fabricator of these first designs, and they claim to have built hundreds of the type all over North America. Illustrated here is Muskegon's "water jacket" burner, the last and most elaborate variation of the type.

The water jacket burner realized a significant reduction in the cost of the structure's refractory lining of fire-brick by surrounding the boiler plate with an eighteen-inch lining of water. This kept the steel temperature low and ingeniously preheated water that could later be used in a mill's powerhouse. Considerably more expensive than earlier versions, however, it served only large, prosperous mills. The only surviving example of the water jacket burner is found in Bonners Ferry, Idaho (fig. 13),

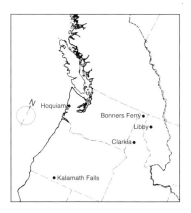

This page:

Fig. 16. Water jacket wood burner, Bonners Ferry, Idaho.

Fig. 17. Diagram of sidewall assembly showing water chamber between plates.

Fig. 18. Map indicating locations of extant cylindrical-type burners.

Opposite:

Fig. 19. Elevation.

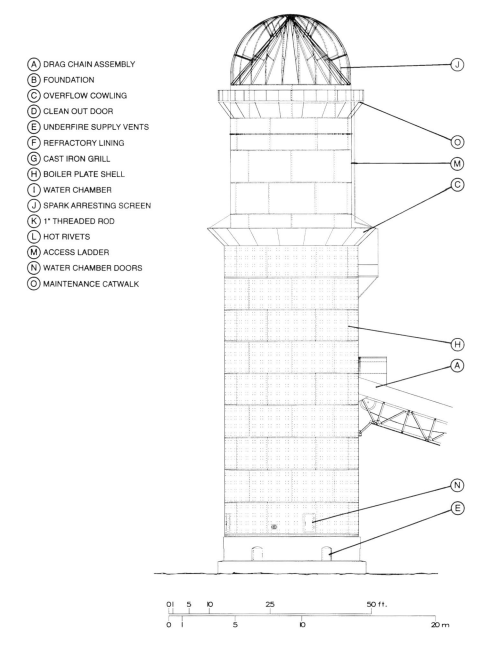

A DRAG CHAIN ASSEMBLY
B FOUNDATION
C OVERFLOW COWLING
D CLEAN OUT DOOR
E UNDERFIRE SUPPLY VENTS
F REFRACTORY LINING
G CAST IRON GRILL
H BOILER PLATE SHELL
I WATER CHAMBER
J SPARK ARRESTING SCREEN
K 1" THREADED ROD
L HOT RIVETS
M ACCESS LADDER
N WATER CHAMBER DOORS
O MAINTENANCE CATWALK

0 1   5   10   25   50 ft.

0 1   5   10   20 m

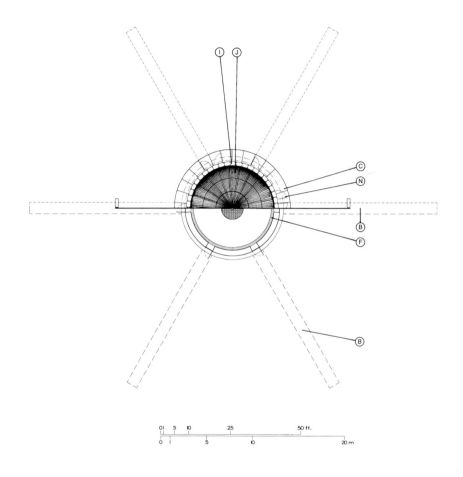

although they were constructed in locations extending from Vera Cruz, Mexico to Portland, Maine.

The version illustrated here is 137 feet high with twelve-by-eight foot sheets of 5/8 inch thick boiler plate for its sidewalls. Figure 17 illustrates a portion of the sidewall assembly. The proportionally large size of the concrete foundation is also of note, as seen in the section (fig. 22).

Cylindrical burners are illustrated in the plates on pages 77–78.

Fig. 20. Axonometric diagram.

Fig. 21. Bilevel plan showing cap (top) and base (bottom) of burner.

Fig. 22. Section.

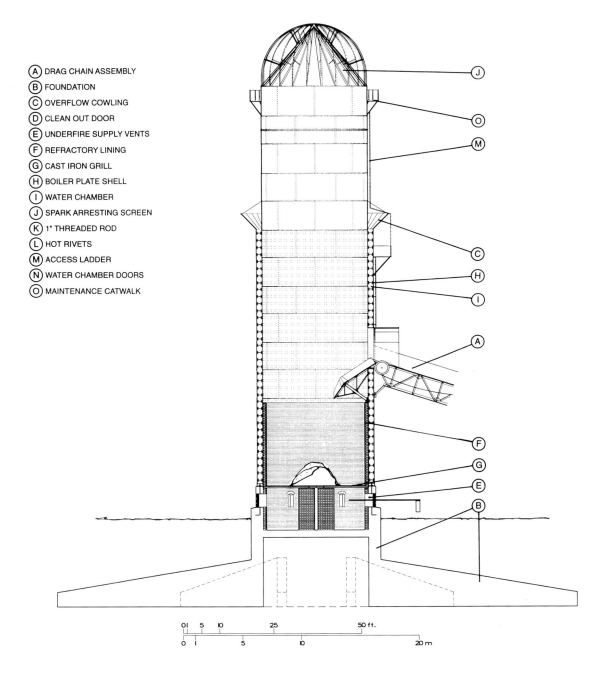

(A) DRAG CHAIN ASSEMBLY
(B) FOUNDATION
(C) OVERFLOW COWLING
(D) CLEAN OUT DOOR
(E) UNDERFIRE SUPPLY VENTS
(F) REFRACTORY LINING
(G) CAST IRON GRILL
(H) BOILER PLATE SHELL
(I) WATER CHAMBER
(J) SPARK ARRESTING SCREEN
(K) 1" THREADED ROD
(L) HOT RIVETS
(M) ACCESS LADDER
(N) WATER CHAMBER DOORS
(O) MAINTENANCE CATWALK

01 5 10   25   50 ft.

0   1   5   10   20 m

29

Colby Engineering's "air-cooled" wood burner of 1916 was the first significant departure from the cylindrical burner type. Although still hefty, it was simpler to construct than previous models and the shell required a far thinner and less expensive gauge of steel. A series of radial trusses separated two layers of cladding between which air flowed, cooling the structure and keeping it safe from collapse (fig. 24). The enlarged conical base also imposed a physical distance between the heat of the fire and the steel shell; the same principal employed in Native American shelters.

Colby's breakthroughs in design inspired other manufacturers and, within a year, a host of similar, competitive models became widely available. Due to its brief period of market popularity, only a few dozen of Colby's burners were constructed, and none are known to stand today. These drawings are derived from the one foundation that is still extant, along with trade journal advertisements, patent information, and historical photographs.

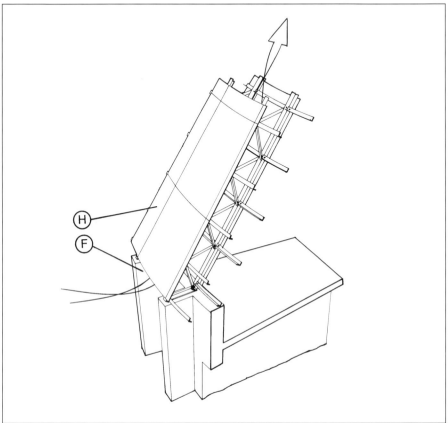

This page:

Fig. 23. Colby air-cooled wood burner, Silverton, Oregon. Photograph from December 1916 advertisement in *West Coast Lumberman*.

Fig. 24. Diagram of sidewall assembly at the burner base indicating direction of air-flow.

Fig. 25. Map indicating location of the remaining air-cooled burner foundations.

Opposite:
Fig. 26. Elevation.

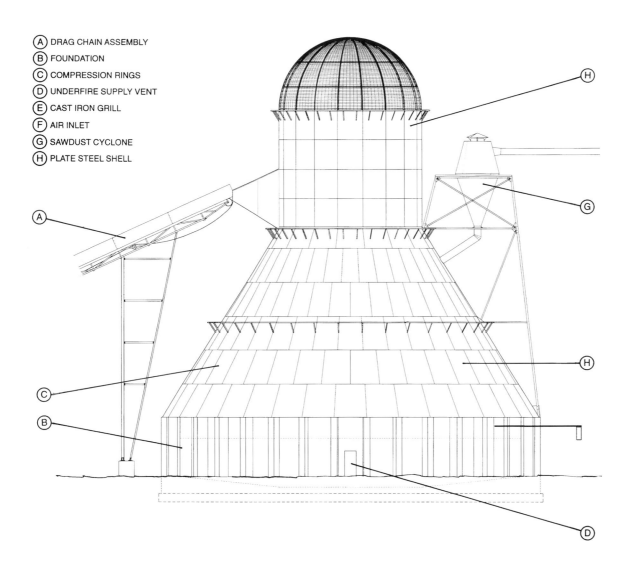

(A) DRAG CHAIN ASSEMBLY
(B) FOUNDATION
(C) COMPRESSION RINGS
(D) UNDERFIRE SUPPLY VENT
(E) CAST IRON GRILL
(F) AIR INLET
(G) SAWDUST CYCLONE
(H) PLATE STEEL SHELL

01 5 10   25   50 ft.

0 1   5   10   20 m

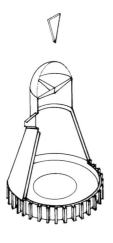

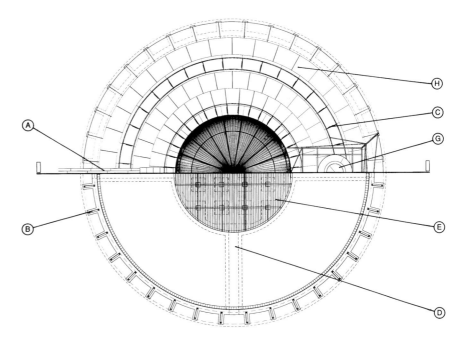

The elevation and section drawings illustrate a cone-shaped device attached to the exterior of the burner, approximately two-thirds from the base to the cap (figs. 26 G & 29 G). Commonly known as "cyclones" or "dust precipitators," they can be found on wood burners of all types. Their function was to funnel planar shavings and small particles of sawdust into the burner. These particles were blown through a pipe into the cyclone from the mill itself. Once inside the cyclone, the chips, dust, and shavings were then drawn by gravity into the core of the burner.

This page:
Fig. 27. Axonometric diagram.

Fig. 28. Bi-level plan showing cap (top) and base (bottom) of burner.

Opposite:
Fig. 29. Section.

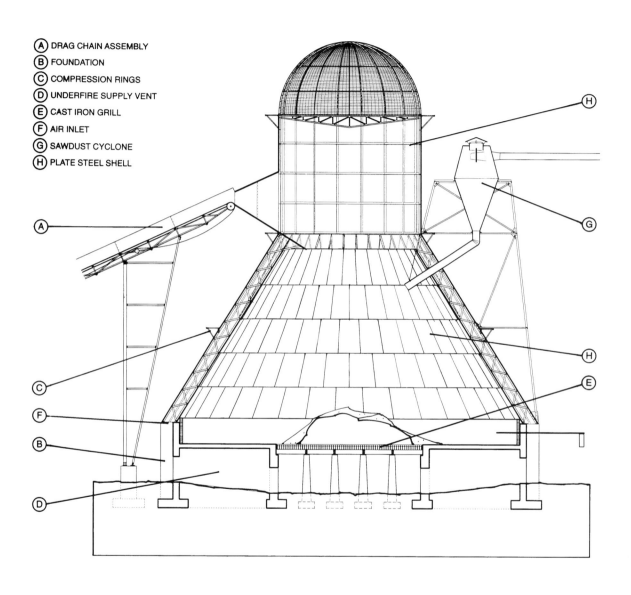

(A) DRAG CHAIN ASSEMBLY
(B) FOUNDATION
(C) COMPRESSION RINGS
(D) UNDERFIRE SUPPLY VENT
(E) CAST IRON GRILL
(F) AIR INLET
(G) SAWDUST CYCLONE
(H) PLATE STEEL SHELL

Single-Shell Burners

Derived from Colby Engineering's "air-cooled" model, this bastardized design commonly appeared at mill sites through the 1950s. It is included here to illustrate the transition from the early air-cooled type burners to the better known "wigwam" variety that would follow. This model is notable for having only a single steel shell—unlike the Colby burner, which had two—a cylindrical top, and a considerably minimized foundation. Of those still extant, two are equipped with dual spark arresting screens, suggesting a relation to earlier, cylindrically designed burners. Although it is unclear who is responsible for this specific design, the number of burners of this type that remain standing suggest that a single fabricator had a proprietary interest in their duplication.

The configuration of the drag chain assembly, which transported the waste material into the core of the burner, is illustrated above (fig. 31). Unlike most drag chain arrangements, which rested on the

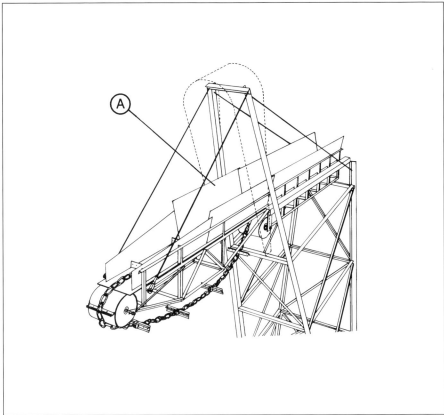

This page:

Fig. 30. Single-shell wood burner in Ronan, Montana.

Fig. 31. Diagram of drag chain assembly used to introduce waste into the burner.

Fig. 32. Map indicating the location of remaining single-shell wood burners.

Opposite:

Fig. 33. Elevation.

A DRAG CHAIN ASSEMBLY
B FOUNDATION
C COMPRESSION RINGS
D CLEAN OUT DOOR
E SACRIFICIAL LINING
F ACCESS LADDER
G SPARK ARRESTING SCREEN
H AIR INLET

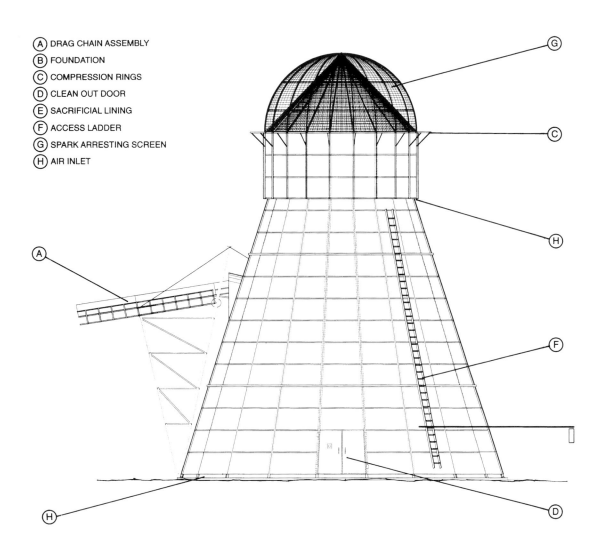

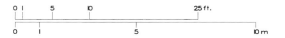

35

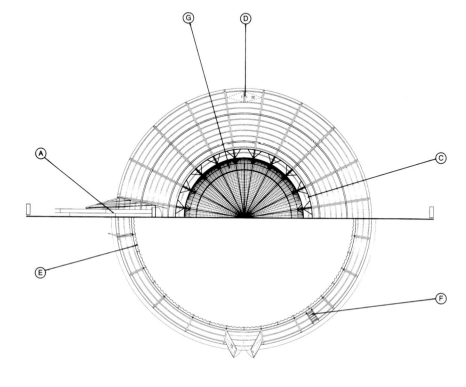

shell, this design ingeniously suspended the assembly over the flames with a system of rods in tension. This kept the structure of the wood burner free of any unnecessary penetrations and loading.

Single-Shell burners are illustrated in the plates on pages 79–81.

This page:
Fig. 34. Axonometric diagram.

Fig. 35. Bi-level plan showing cap (top) and base (bottom) of burner.

Opposite:
Fig. 36. Section.

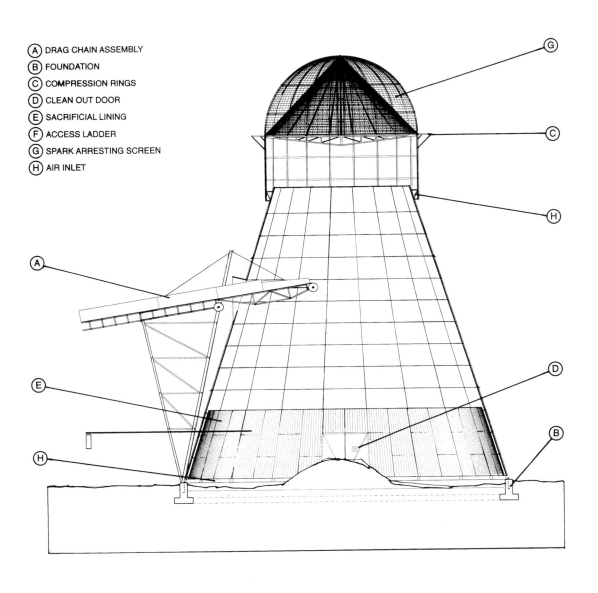

- Ⓐ DRAG CHAIN ASSEMBLY
- Ⓑ FOUNDATION
- Ⓒ COMPRESSION RINGS
- Ⓓ CLEAN OUT DOOR
- Ⓔ SACRIFICIAL LINING
- Ⓕ ACCESS LADDER
- Ⓖ SPARK ARRESTING SCREEN
- Ⓗ AIR INLET

The "Wigwam" or "Beehive" Burner

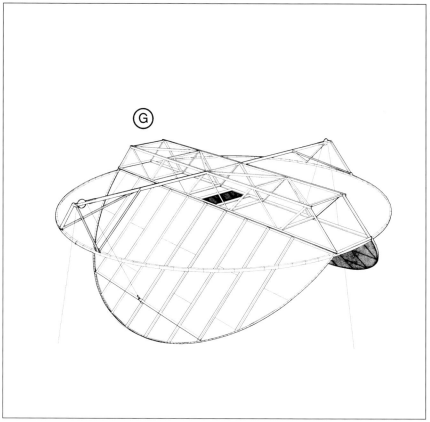

The following drawings represent the most common wood burner type, known colloquially as the "wigwam" burner in America, and as the "beehive" burner in Canada. From 1917 through the 1970s, thousands of this type of incinerator were constructed throughout the Pacific Northwest by dozens of independent fabricators. Common to all of these designs were spark arresting screens, cyclone vents, maintenance catwalks, compression rings, and large doors for cleaning purposes.

In response to the provisions of the 1970 Clean Air Act, American versions of this basic design were often modified to accommodate pollution control measures. Underfire fan units (figs. 40 I & 43 I) and downward acting, semicircular damper lids (figs. 40 G & 43 G) were introduced on many of these burners, in order to increase the combustion chamber temperature to approximately eight hundred degrees Fahrenheit, and thereby allow for more efficient incineration.

This page:

Fig. 37. Wigwam burner in operation, Donald, British Columbia.

Fig. 38. Diagram of temperature controlling damper lid.

Fig. 39. Map indicating the location of well-preserved wigwam and beehive burners.

Opposite:
Fig. 40. Elevation.

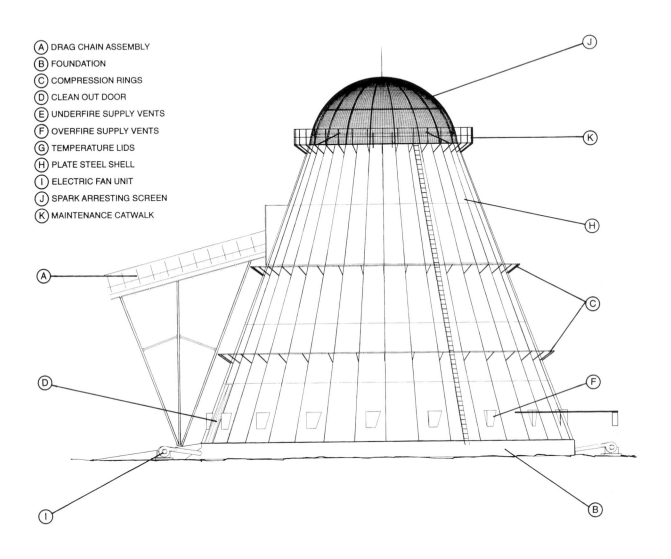

(A) DRAG CHAIN ASSEMBLY
(B) FOUNDATION
(C) COMPRESSION RINGS
(D) CLEAN OUT DOOR
(E) UNDERFIRE SUPPLY VENTS
(F) OVERFIRE SUPPLY VENTS
(G) TEMPERATURE LIDS
(H) PLATE STEEL SHELL
(I) ELECTRIC FAN UNIT
(J) SPARK ARRESTING SCREEN
(K) MAINTENANCE CATWALK

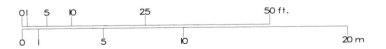

0 1   5   10   25   50 ft.

0   1   5   10   20 m

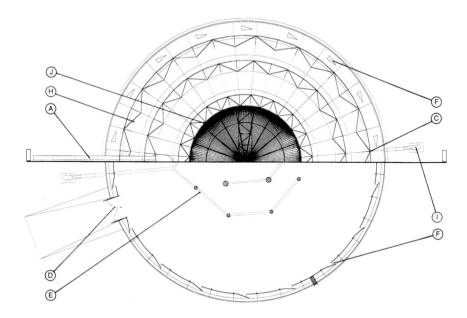

The damper lid assembly (fig. 38) could be pre-fabricated off-site and installed within one-and-a-half days, thereby minimally affecting the production of the mill. The map (fig. 39) indicates the locations of the best preserved and most interesting examples of this type. It does not include more than fifty unremarkable and or badly deteriorated burners.

Wigwam and beehive burners are illustrated in the plates on pages 82–103.

This page:
Fig. 41. Axonometric diagram.

Fig. 42. Bi-level plan showing cap (top) and base (bottom) of burner.

Opposite:
Fig. 43. Section.

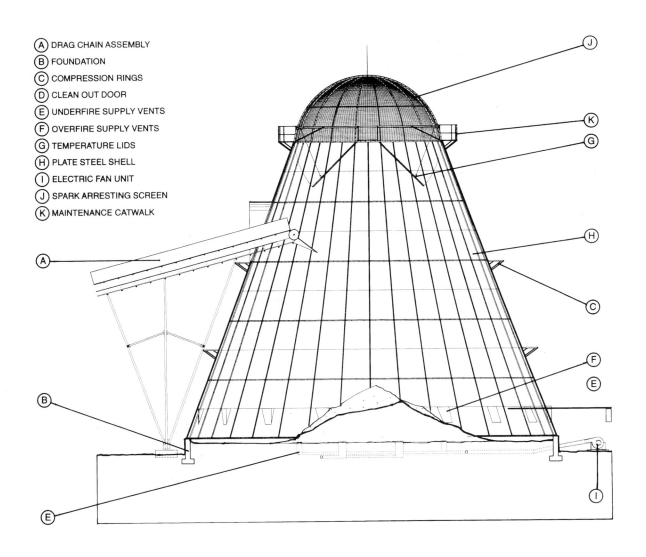

- Ⓐ DRAG CHAIN ASSEMBLY
- Ⓑ FOUNDATION
- Ⓒ COMPRESSION RINGS
- Ⓓ CLEAN OUT DOOR
- Ⓔ UNDERFIRE SUPPLY VENTS
- Ⓕ OVERFIRE SUPPLY VENTS
- Ⓖ TEMPERATURE LIDS
- Ⓗ PLATE STEEL SHELL
- Ⓘ ELECTRIC FAN UNIT
- Ⓙ SPARK ARRESTING SCREEN
- Ⓚ MAINTENANCE CATWALK

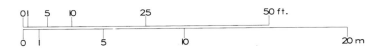

01 5 10   25   50 ft.

0 1   5   10   20 m

41

The Canadian Smokeless Burner

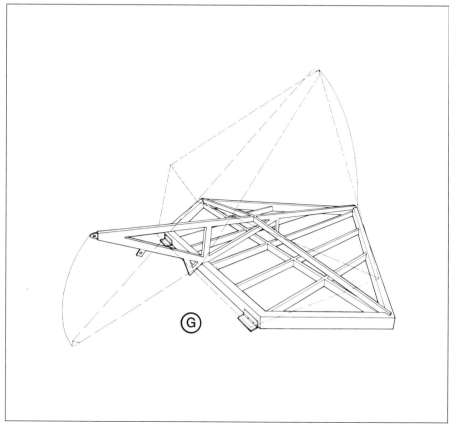

The Canadian version of the smokeless wood burner type developed in response to the increasingly stringent air quality standards imposed on urban areas of British Columbia and Alberta in the mid-1970s. By 1980, virtually all new wood burners being constructed were of this type. Better fan units for the underfire and overfire vents, as well as a superior damper lid design, allowed these burners to operate at significantly higher temperatures (between eight hundred and twelve hundred degrees Fahrenheit) than previous possible. Under optimal conditions, the Canadian smokeless exhibited no visible signs of emission.

The earliest versions of this type can be distinguished by the absence of the dome-shaped spark arresting screens at their caps; it was thought that the higher temperatures at which these burners operated would prevent the release of sparks, making the screen obsolete. However, when wet or impure waste was burned, the smoke, sparks, and sooty particles

This page:

Fig. 44. Smokeless wood burner, Fort Fraser, British Columbia.

Fig. 45. Diagram of upward-acting damper lid leaf.

Fig. 46. Map indicating the location of towns with well preserved or numerous examples of the Canadian smokeless type.

Opposite:
Fig. 47. Elevation.

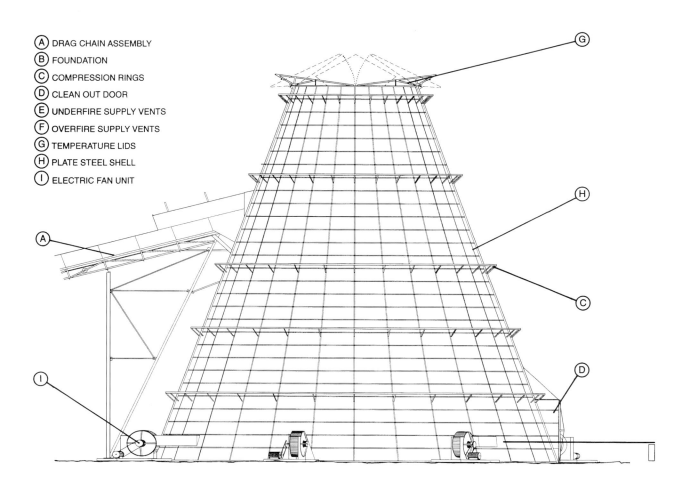

(A) DRAG CHAIN ASSEMBLY
(B) FOUNDATION
(C) COMPRESSION RINGS
(D) CLEAN OUT DOOR
(E) UNDERFIRE SUPPLY VENTS
(F) OVERFIRE SUPPLY VENTS
(G) TEMPERATURE LIDS
(H) PLATE STEEL SHELL
(I) ELECTRIC FAN UNIT

01 5 10   25   50 ft.

0 1   5   10   20 m

returned, resulting in the reintroduction of the screens.

One leaf of the basic configuration of the upward acting damper lids is illustrated here (fig. 45). The location map indicates towns with particularly good or numerous examples of this type of burner (fig. 48). More than thirty locations with unremarkable or damaged burners are not indicated.

Canadian smokeless burners are illustrated in the plates on pages 110–115.

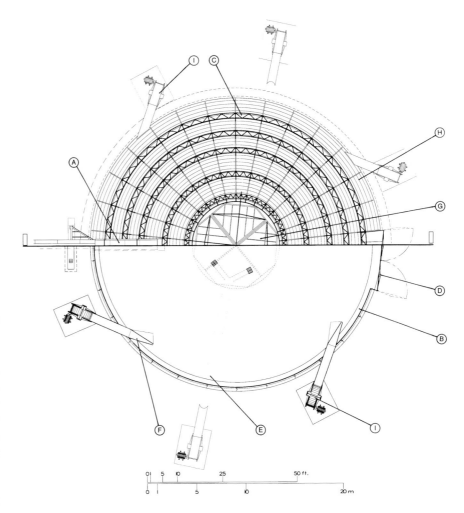

This page:
Fig. 48. Axonometric diagram.

Fig. 49. Bi-level plan showing cap (top) and base (bottom) of burner.

Opposite:
Fig. 50. Section.

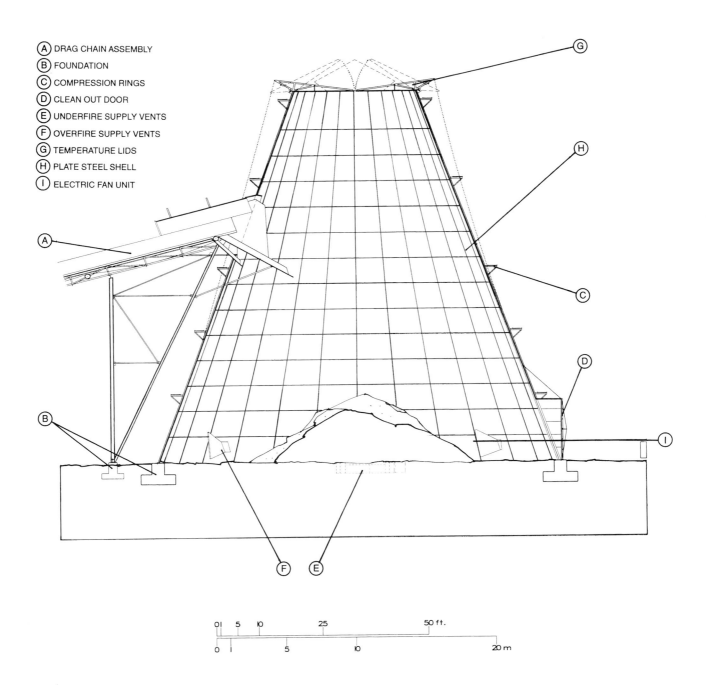

A DRAG CHAIN ASSEMBLY
B FOUNDATION
C COMPRESSION RINGS
D CLEAN OUT DOOR
E UNDERFIRE SUPPLY VENTS
F OVERFIRE SUPPLY VENTS
G TEMPERATURE LIDS
H PLATE STEEL SHELL
I ELECTRIC FAN UNIT

01 5 10 25 50 ft.

0 1 5 10 20 m

INTERIOR PLATES

These

are the desolate, dark weeks
when nature in its barrenness
equals the stupidity of man.

The year plunges into night
and the heart plunges
lower than night

To an empty, windswept place
without sun, stars or moon
but a peculiar light as of thought

that spins a dark fire—
whirling upon itself until,
in the cold, it kindles

—*William Carlos Williams*

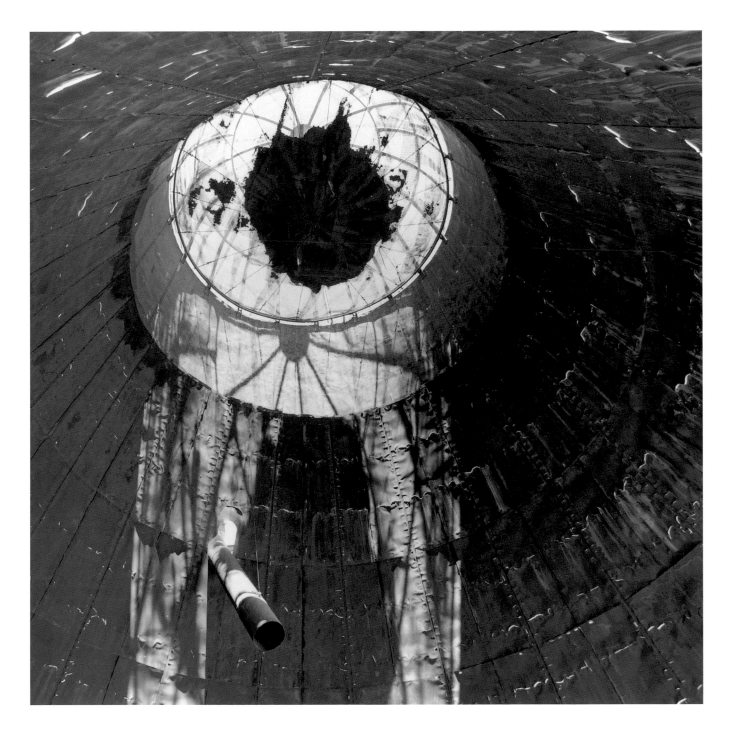

Interior location key

EXTERIOR PLATES

"Vrindaban"

Everything sculptured
                from color to form
from form to fire
                Everything was vanishing
Music of wood and metal
in the cell of the god
                womb of the temple

—*Octavio Paz*

Exterior location key

74	Knutsford, British Columbia
75	Clarkia, Idaho
76	Hoquiam, Washington
77	Kalamath Falls, Oregon
78	Bonners Ferry, Idaho
79	Nubieber, California
80	Ronan, Montana
81	Colville, Washington
82	Aidin, California
83	Seneca, Oregon
84	Forks, Washington
85	Oakland, Oregon
86	Sattley, California
87	Alturas, California
88	Smithers, British Columbia
89	Forks, Washington
90	Bonners Ferry, Idaho
91	Sweet Home, Oregon
92	Trout Creek, Montana
93	Donald, British Columbia
94	Thompson Falls, British Columbia

95	Gilchrist, Oregon
96	Smithers, British Columbia
97	Merritt, British Columbia
98	Missoula, Montana
99	Darby, Montana
100	Salmon, Idaho
101	Thompson Falls, British Columbia
102	Wieppe, Idaho
103	Cranbrook, British Columbia
104	Deerlodge, Montana
105	Marysville, California
106	Vancouver, British Columbia
107	Copalis Crossing, Washington
108	Boville, Idaho
109	Astoria, Oregon
110	Smithers, British Columbia
111	Lumby, British Columbia
112	Penticton, British Columbia
113	Oakanogan Falls, British Columbia
114	Cranbrook, British Columbia
115	Malakwa, British Columbia

Selected Bibliography

Andrews, Ralph W. *This Was Sawmilling*. Seattle: Superior Publishing Company, 1957.

Atherton, George H., Russell W. Bonlie, Stanley E. Corder, and Paul E. Hyde. "Wood and Bark Residue Disposal in Wigwam Burners." *Bulletin*, Forest Research Laboratory, Oregon State University, no. 11, paper 720, (March 1970).

Clay, Grady. "Ephemeral Places: Here Today Gone Tomorrow." *Design Quarterly*, 143, edited by Mildred Friedman. Cambridge, MA: MIT Press, 1989.

Cox, Thomas R. *Mills and Markets: A History of the Pacific Coast Lumber Industry to 1900*. Seattle: University of Washington Press, 1974.

Eisenman, Peter. "Introduction." In *The Architecture of the City*, by Aldo Rossi. Cambridge, MA: MIT Press, 1982.

Glassie, Henry. *Folk Housing in Middle Virginia*. Knoxville: University of Tennessee Press, 1975.

Holl, Steven. *Intertwining: Selected Projects 1989-1995*. New York: Princeton Architectural Press, 1996.

Holl, Steven, Juhani Pallasmaa, and Alberto Pérez-Gómez. *Questions of Perception: Phenomenology of Architecture*. Tokyo: A + U Publishing Co., July 1994 (special issue).

Lucia, Ellis. *The Big Woods: Logging and Lumbering, From Bull Teams to Helicopters in the Pacific Northwest*. Garden City, NY: Doubleday, 1975.

Marx, Leo. *The Machine in the Garden: Technology and the Pastoral Ideal in America*. New York: Oxford University Press, 1964.

Nabakov, Peter. *Native American Architecture*. New York: Oxford University Press, 1989.

Nesbitt, Kate, ed. *Theorizing a New Agenda for Architecture: An Anthology of Architectural Theory 1965–1995*. New York: Princeton Architectural Press, 1996.

Norburg-Schulz, Christian. *Existence, Space & Architecture*. New York: Praeger Publishers, Inc., 1971.

Oakleaf, H. B. *Lumber Manufacturing in the Douglas Fir Region*. Chicago: Commercial Journal Co., 1920.

Rossi, Aldo. *The Architecture of the City*. Cambridge, MA: MIT Press, 1982.

Style and Vernacular: A Guide to the Architecture of Lane County, Oregon. Portland, OR: Western Imprints, 1983.

Vaughn, Thomas, ed. *Space, Style and Structure*. Portland, OR: Oregon Historical Society, 1964.

Epilog

"Time flows through these places, leaving lessons along the paths. But not all places exhibit the same viscosity: time and life move more quickly through some than through others, extending their influence. . . . You will not find them on city maps. But these special locales carry huge layers of symbolism and meaning. Such geographic gestures have the capacity to package emotion, energy, history, and allusion into compact space. Here one sees a larger place and its character writ small, in compression or miniature. They stand for other things, they put it all together."

—Grady Clay, *Ephemeral Places*

The memory of every wood burner is a node in a fibrous net that transparently envelops the region, connecting hundreds of small towns. As symbols, burners offers many contradictory readings. One might see the transformation of the forest into ash and plumes of smoke, or, conversely, think of them as the heroic repositories of human labor. Others might interpret them as symbols for the resiliency of nature, the negligence of man, the end of an era, the hope of the future, or of the hopelessly nostalgic. With its function eliminated, the wood burner is now adrift: past, present, and future are all tenses with which it collaborates.

As an object that can be dismantled and reassembled anywhere, an icon for a region in flux, a construction of materials that age, and a vessel that captures light, the wood burner is the embodiment of the concept of change.

Beyond these individual readings, any attempt at an entirely intellectual or categorical determination of their meaning would be futile. As Alberto Pérez-Gómez has written, ". . . the meaning of the work lies in the fact that it is there. . . . It is, first and foremost, of the world, and our experience of it overwhelms us."

In the absence of such closure, we find that we are drawn to the wood burner by its powerful physical and phenomenal presence, yet simultaneously repelled by its history of rapacious consumption. By suspending these two thoughts in a dialectic exchange, perhaps we can understand its generative potential and cultural significance more rationally, without the distorting lenses of history and nostalgia. It is hoped that such an understanding will create a small niche in the catalogs of history for these unique and powerful structures, despite their likely disappearance in this complex era of receding cultures and dwindling resources.

Acknowledgements

In the six years that have elapsed from the germination of an idea through the production of this text I have received support and assistance from many wonderful individuals and organizations.

Financial assistance is a key ingredient in the completion of works of this nature. The principle supporters of this project, in order of receipt, have been: the Louis C. Rosenburg Scholarship, offered through the University of Oregon School of Architecture and Allied Arts, the Graham Foundation for Advanced Studies in the Arts, and the Betty Bowen Committee of the Seattle Art Museum.

I was extremely fortunate to have had my dear friend Neil Chowdhury, a skilled photographer, accompany me on on my first extended sawmill expedition. Neil's patience, generosity, and talent has granted this work strengths it would not otherwise have had. His spirit is everywhere in this book.

Steven Holl has been a catalytic force in making this book a reality. I am heavily indebted to him for the inspiration to make this a serious work, encouragement to carry it through, letters of support, critical advice, and the contribution of a beautifully written foreword to this volume.

At the University of Oregon, I had the unrelenting support of Professors Peter Keyes and John Reynolds. Professors John Cava, Brad Cloepfil, Jerry Finrow, Wayne Jewitt, Rob Peña, Don Peting, and Jim Tice also extended themselves for the benefit of this project. During the final preparation of this manuscript, I was very thankful to have had the clear editorial craftsmanship of Cameron Hall.

The Miller/Hull Partnership exercised a great deal of clemency in allowing me not only the time away necessary to complete the manuscript, but also the free and generous use of their facilities, for which I am grateful.

The staff of Princeton Architectural Press has been especially supportive and has provided a greatly needed vote of confidence in the completion of this work. Mark Lamster, Kevin Lippert, and Clare Jacobson were instrumental on this behalf.

My sincere gratitude and appreciation goes out to the following people: Christen Bailey, Molly Blieden, Bill Brewer and the Brewer family, John Cava, Joseph Chaijaroen, Patrick and Lucy Ellison, Kirsten Finstad, Heidi Froehring, Gabe Hajiani, Sean Hall and the Hall family, the Harsey family, Frank Hopkins and Seattle Boiler Works, Kyle Lommen, Sharon Genasci, Lisa Mahar-Keplinger, Corey Martin, Peter Mavridis, Lewis L. McArthur Jr., Jeff Mihalyo, Tom and Kathleen Morris, the Palevitch family, the Schultz family, Gary Snyder and the Snyder family, Leigh Stevens and Muskegon Boiler Works, John Thompson, the Tuttle family, Rebecca Wee, Randy Walker, and to my parents John and Andrée Mihalyo. My sincere gratitude also goes out to the many people in the forest industry who assisted me along the way.

Lastly, but most dearly, my faithful supporter Annie Han has assisted me in ways really too numerous to describe. It should be said, however, that her skill is evidenced in many of the drawings in this volume, and her valuable criticism has guided me throughout this project.

ASSEMBLERS & ERECTORS OF
REFUSE BURNERS.